# ARTEMIS 1 LAUNCH

*NASA Mission To Return
Humans Back To The Moon*

**Sydney Walker**

# Table Of Contents

# INTRODUCTION

Five decades after the last flight of NASA's famed Saturn 5 moon rocket, the U.S. space agency is preparing to launch its most powerful rocket ever Monday for a key, long-overdue test mission, sending an unpiloted Orion crew capsule on a 42-day cruise around the moon. Running years behind schedule and billions over budget, the first Space Launch System (SLS) rocket is finally poised for blastoff from pad 39B at the Kennedy Space Center at 8:33 a.m. EDT Monday, the start of a two-hour window. Forecasters are expecting a 70 percent likelihood of nice weather. Backup launch possibilities are possible on September 2 and 5 depending on the projected trajectory and the ever-changing locations of the Earth and moon. After that, the trip presumably would slide into October. Cobbled together from left-over space shuttle components, a new core stage, and a modified upper stage borrowed from another rocket, the SLS rocket stands 322 feet tall and will weigh 5.75 million pounds after 750,000 gallons of super cold liquid oxygen and hydrogen rocket fuel are pumped aboard early Monday. At liftoff, the SLS will generate a ground-shaking 8.8 million pounds of thrust from four shuttle-era hydrogen-fueled engines and twin solid rocket boosters packed with 25% more propellant than their shuttle predecessors, providing a breathtaking spectacle for thousands of spaceport workers, area residents, and tourists. It is a high-wire act-up. This is a major thing.

And it is gorgeous. And it is a monster! The immensity simply overwhelms you. The major purpose of the Artemis 1 mission is to deploy Orion to orbit around the moon and in the process, set up a 25,000-mph plummet down into Earth's atmosphere on October 10. The major aim of the mission is to check that the capsule's 16.5-foot-wide heat shield can protect returning astronauts from the 5,000-degree inferno of re-entry on a future journey. "This is a test flight. It's not without danger," Bob Cabana, a former shuttle captain and now a NASA associate administrator, said of the maiden SLS launch. "We have studied the risk as best we can and we've reduced it as best we can. But we are taxing Orion beyond what it was truly built for in preparation for delivering it to the moon with a crew. And we want to make sure everything works exactly properly when we do it and that we understand all the hazards," he added. "We're going to learn a lot from this test flight. If the unpiloted Artemis 1 test mission goes well, NASA aims to fly four humans aboard the second SLS rocket for an around-the-moon shakedown trip in 2024 Artemis 2 before the first woman and the first person of color touch down near the moon's south pole in 2025 or 2026. After then, NASA aims to start a constant stream of Artemis lunar missions, sending people to the south pole zone once per year or so for study and to look for ice deposits in permanently shadowed craters, a resource future crews may convert into rocket fuel, air, and water. But first, the Artemis crew and spacecraft have to get there. And it needs a rocket capable of blasting the men, women, and machines out of Earth's gravitational clutches and over the 240,000-mile gulf to the moon with adequate fuel, food, and equipment to undertake a meaningful mission and send the crew safely home when it's done. "She is an excellent rocket," Charlie Blackwell-Thompson, NASA's first female launch director, told CBS News. "She provides a whole new capacity to our nation's space program, a new heavy lift capability for deep space exploration. It's going to revolutionize the method by which we investigate. It's going to restore our country to the moon, and it is going to lay the path for our next steps as we prepare to travel to somewhere like Mars and even places beyond. The original 322-foot SLS "block 1" version can carry 95 tons of cargo and fuel to low-Earth orbit and can transport 27 tons onto the moon. It is the only rocket in the world that can blast that much stuff to the moon in a single trip and it is the only heavy lifter that is already human-rated. Future block 1B and 2 variations, the former employing a more powerful four-engine Exploration Upper Stage and the latter using both the EUS and more powerful boosters, will stand more than 350 feet tall and be capable of carrying between 38 and 47 tons of cargo to the moon. A gigantic rocket from SpaceX. But the SLS is not the only giant rocket presently under development. SpaceX is creating an even more ambitious rocket, one that dwarfs the SLS and everything else on the planning board: a reusable two-stage monster dubbed the Super Heavy-Starship. The Super Heavy first stage will generate a record 16 million pounds of thrust from 33 methane-burning Raptor engines while the Starship upper stage, equipped with six Raptors, life support systems, and crew accommodations, is designed to carry passengers and cargo to the moon and beyond on NASA-sponsored flights or purely commercial ventures. SpaceX estimates the 394-foot-tall 30-foot-wide rocket will be able to send 100 tons or more to the moon, double the capabilities of even the SLS Block 2. But the Super Heavy-Starship can't achieve it in a single trip. Multiple launches of Starship tankers will be necessary to refuel moon-bound spacecraft before they leave Earth orbit and a large delay or launch catastrophe might have substantial ramifications. SpaceX's Super Heavy-Starship rocket undergoing testing at the company's Boca Chica, Texas, flight test site on the Gulf of Mexico. The reusable SpaceX rocket is more powerful than NASA's disposable Space Launch System moon rocket, but it needs refueling in low-Earth orbit. While less costly than the SLS, it's not yet known when the vehicle would be ready for operational usage. No government or corporation has ever carried out orbital refueling on such a vast scale and it's a capacity SpaceX has yet to show. But Musk is convinced the technology will function. SpaceX already is constructing a Starship derivative to serve as NASA's first Artemis lunar lander under a $2.9 billion contract, and the ability to refuel the spacecraft in Earth orbit will be essential. Orion is constructed as a deep space exploration vessel, SLS is supposed to transport it there. That's what SLS does," said Jim Free, NASA's director of exploration systems. "SpaceX is a partner (and) we buy into what SpaceX is attempting to achieve. But right now, they don't have the capability that SLS does. The SpaceX Super Heavy-Starship has one big advantage over the government-managed, owned, and controlled SLS: affordability. While SpaceX does not share development costs, the Super Heavy Starship is estimated to be orders of magnitude less costly than the SLS. According to NASA's Inspector General, the U.S. space agency is scheduled to spend $93 billion on the Artemis (moon program) up to FY 2025. We also forecast the current manufacturing and operating cost of a single SLS/Orion system at $4.1 billion per launch for Artemis 1 through 4, however, the Agency's continuing measures focused on boosting affordability strive to minimize that cost. Among the causes listed as contributing to the SLS's astronomical price tag: are the use of sole-source, cost-plus contracts and the fact that except for the Orion capsule, its subsystems, and the supporting launch facilities, all components are expendable and 'single use' unlike emerging commercial space flight systems.

EUROPEAN SPACE AGENCY

In sharp contrast to SpaceX's dedication to entirely reusable rockets, everything that saves the Orion crew capsule is destroyed after a single use. As SpaceX founder Musk likes to point out, that's like flying a 747 jumbo aircraft from New York to Los Angeles and then tossing the airplane away. This is an expendable, single-use system unlike some of the launch systems that are out there on the commercial side of the business, where there are several uses. This is a single-use system. And so the $4.1 billion per trip, bothers us enough that in our reports, we stated we regard it as unsustainable. But the SLS has one evident near-term advantage: flight-tested components. When it approved the SLS project at the end of the space shuttle program, Congress required NASA to use available hardware if possible. The SLS Block 1 uses modified shuttle-heritage main engines and a Northrop Grumman booster system that is already human rated the

Artemis 1 engines have flown a combined 25 shuttle flights along with a Boeing-designed upper stage that's used with United Launch Alliance's Delta 4 rocket. Even Orion's European Space Agency-supplied service module, built by Airbus, has flight heritage. Its main thruster is a repurposed space shuttle Orbital Maneuvering System engine, built by Aerojet Rocketdyne, that flew 19 times between 1984 and 2002. As for the exorbitant cost, Marcia Smith, a Washington-based space expert, noted in an email exchange that money isn't necessarily the most essential aspect. For SLS, protecting employment, not just jobs per se, but high-tech jobs in a field crucial for national security is a key motive. If, as a country, it is vital to lead the world in space exploration, do you want to put all of your eggs in the billionaire space enthusiast's basket? Bet your everything on individuals who may alter their minds and go away or suffer disease or worse? They are single-point failures. If the SLS experiences a catastrophic failure, "the narrative might change," she noted. "But even then I'm not sure. Not everyone is confident that the private sector is dependable enough to risk the nation's space leadership on public-private partnerships."

# CHAPTER 1

## *ABOUT THE MISSION*

All eyes will be on the iconic Launch Complex 39B when the Orion spacecraft and the Space Launch System (SLS) rocket take off for the first time from NASA's refurbished Kennedy Space Center in Florida. Artemis I will be the first in a succession of progressively sophisticated missions to construct a long-term human presence on the Moon for decades to come. The key aims for Artemis I are to demonstrate Orion's systems in a spaceflight environment and assure a safe re-entry, descent, splashdown, and recovery before the first flight with the crew on Artemis II.

## *Mission Facts:*

- **Launch Date**: Aug. 29, 2022
- **Mission Duration**: 42 days, 3 hours, 20 minutes
- **Total Distance Traveled**: 1.3 million miles
- **Re-entry Speed**: 24,500 mph (Mach 32) (Mach 32)
- **Splashdown**: Oct. 10, 2022

Launch Preparations Remain on Track, Weather 70% Favorable. Engineers continue to prepare NASA's giant Moon rocket, Orion spacecraft, and ground systems for the Artemis I launch. The two-hour launch window starts at 8:33 a.m. EDT on Monday, Aug. 29. At Launch Pad 39B, workers finished repairing the hydraulic power units on the Space Launch System rocket's boosters Wednesday. After...

## *The Mission Patch For Artemis I*

The mission patch for Artemis I highlights various components within the design that provide significant value for this historic trip including the triangle form and the colors silver, orange, red, white, and blue.

## *SPACE LAUNCH SYSTEM ROCKET*

The most powerful rocket in the world, intended to transport people to deep space.

- HEIGHT : 322 feet
- MASS AT LIFTOFF: 5.75 million pounds\sTHRUST AT LIFTOFF: 8.8 million pounds
- PAYLOAD TO THE MOON: 59,000 pounds

- ORION SPACECRAFT: Next generation spacecraft, built for the challenges of human trips to deep space.

- CREW AND SERVICE MODULE HEIGHT: 26 feet
- PRESSURIZED VOLUME: 690.6 ft 3\s
- MASS TO THE MOON: 53,000 pounds
- RETURN MASS AT LANDING: 18,200 pounds

- BUILDING AND TESTING: Every state in America has contributed to developing Artemis, with firms hard at work to create the technologies that will assist establish a long-term human presence on the Moon. Contributions from men and women throughout America and Europe are important to the space economy, generating new businesses and technologies, supporting employment development, and boosting the need for a highly educated workforce.

- FOLLOW OUR PROGRESS: It's an exhilarating excursion into the expanse of outer space. Follow along and we'll take you there.

# CHAPTER 2

## *WHAT WILL ARTEMIS 1 DO?*

NASA's Artemis 1 mission will launch on the first Space Launch System rocket and its Orion capsule on Monday, Aug. 29. Artemis 1 mission will inaugurate a new era of U.S. space research when it launches to the moon this month, but precisely when it takes off depends on numerous things. Artemis 1, the first uncrewed test flight of NASA's Artemis program to return people to the moon, is presently slated to blast off from Pad 39B of the Kennedy Space Center in Cape Canaveral on Monday (Aug. 29). (Aug. 29). Liftoff is presently slated for 8:33 a.m. EDT (1233 GMT), weather permitting. You may watch the launch live online on Monday beginning at 6:30 a.m. EDT (1030 GMT), courtesy of NASA TV. NASA has emphasised repeatedly that Artemis 1 is, at its heart, a test trip. It marks the first-ever launch of the agency's new mega-rocket, the towering Space Launch System (SLS), as well as the first deep-space voyage for the new Orion spacecraft. There may be technical difficulties that crop up during the launch countdown that require a delay. "The test combat itself has inherent danger," Jim Free, NASA assistant administrator for exploration systems development, said in an Aug. 22 press conference. "This is the first launch of a new rocket and a new spaceship." NASA has a two-hour window in which to launch Artemis 1 on Aug. 29. Which implies the launch may occur anywhere between 8:33 a.m. and 10:33 a.m. EDT (1233-1433 GMT), but NASA is targeting the opening of the window. The weather might potentially cause a delay. Currently, there is a 70% likelihood of excellent weather at launch time, according to an Aug. 25 forecast(opens in new tab) from the Space Launch Delta 45 weather group at Cape Canaveral Space Force Station. The biggest concerns are dense clouds, surface electrical fields from lightning, and the likelihood the SLS may have to fly through the rain on its climb. The weather group will offer daily weather reports until launch.

## *WHAT IF ARTEMIS 1 DOESN'T LAUNCH ON AUG. 29?*

You can launch a Space Launch System of your own with this Estes NASA SLS model rocket(opens in new tab) for a 1:200 size recreation of NASA's moon mega-rocket. Read more about it. If technical challenges or poor weather delay the Artemis 1 mission, NASA does have several possibilities. There are two backup launch days in Artemis 1's current flight window: Friday, Sept. 2, and Monday, Sept. 5. Both dates have their extended launch windows. If NASA is compelled to go for the Sept. 2 launch date, the Artemis 1 SLS rocket would launch at

12:48 p.m. EDT (1748 GMT) and would have a two-hour window to get off the ground. NASA would have to accept a shorter mission, 39 days instead of the 42-day journey an Aug. 29 liftoff allows for if the agency opts for this launch date. It would splash down in the water on Oct. 11 instead of the scheduled Oct. 10. The Sept. 5 launch date calls for liftoff at 5:12 p.m. EDT (2212 GMT) (2212 GMT). The launch window for this date is a little shorter, 90 minutes as opposed to two hours, but does enable NASA to attempt a longer 42-day voyage. Landing would occur on Oct. 17.

## *WHAT IF NASA MISSES THIS ARTEMIS 1 LAUNCH WINDOW?*

If NASA is unable to launch the Artemis 1 mission during the Aug. 29 to Sept. 5 timeframe, the agency would have to fall back on a series of subsequent launch windows that span throughout the remainder of the year and early 2023. In May, the space agency published a calendar of launch options until mid-2023 that satisfied a range of requirements required for the Artemis 1 mission. That list was revised earlier this month as NASA examined various launch possibilities.

Here's a look at when those extra launch possibilities may arise.

Sept. 19-Oct. 4, except for Sept. 29-30;\sOct. 17-Oct. 31, except for Oct. 24-26 and Oct. 28;\sNov. 12-Nov. 27, except for Nov. 20-21 and Nov. 26 (preliminary);\sDec. 9-23, except for Dec. 10, 14, 18 and 23 (preliminary) (preliminary).

# *HOW DOES NASA PICK ARTEMIS 1 LAUNCH DATES?*

NASA has four major considerations that influence how it picks launch dates for Artemis 1. Here's what they are, according to a NASA datasheet:

- The moon's position in its orbit: The moon's location influences the launch day because NASA intends to deploy Artemis 1 in what's termed a "distant retrograde orbit" around the moon. To achieve so, the SLS current upper stage needs to undertake a maneuver called a trans-lunar injection burn at a certain moment, about the moon's location, to set it on the appropriate trajectory.

- Orion's illumination conditions: NASA mission standards stipulate that the Orion spacecraft cannot remain in darkness for more than 90 minutes at a time. This is because Orion is solar powered, therefore it requires sunshine on its arrays, and also to maintain its ideal temperature.

- Orion's "skip" reentry plan: To localize Orion's splashdown site in the Pacific Ocean, NASA plans to perform a "skip" reentry in which the space capsule dips into the Earth's upper atmosphere, then skips out temporarily before completing its final reentry. To achieve that, it must launch within specified windows to attain the appropriate trajectory upon its return to Earth,

- NASA says. Orion's splashdown illumination conditions: NASA wants Orion to splash down during the daylight hours to make it simpler for recovery workers to identify the spacecraft and pull it out of the water.

# CHAPTER 3

## *HOW LONG IS ARTEMIS 1 MISSION?*

As the United States prepares to once again place men on the lunar surface, other nations are eager to explore the moon, too, with rovers and landers. The uncrewed test mission represents phase one of the space agency's historic return to the moon. The Space Launch System Rocket on the pad on August 17 NASA / Joel Kowsky. NASA's new moon mission is ready to bust all sorts of records for human spaceflight. Named after the Greek goddess Artemis, Apollo's twin sister, this endeavor will send the first woman and first person of color to the moon. If everything goes as expected, in 2025, these astronauts will become the first humans to foot on the lunar regolith—or dusty moon soil—since Apollo 17's Gene Cernan and Harrison Schmitt stepped there in December 1972. In addition, the Artemis mission will create the first long-term human presence on the moon, by establishing a space station in orbit and erecting a base camp on the lunar surface. These efforts will establish the framework for yet another first in the future: sending humans to Mars. But before all that occurs, the space agency needs to test its equipment with a voyage named Artemis 1, which will shatter records of its own. As NASA's huge Space Launch System (SLS) rocket stands on the launch pad ahead of this historic mission, here's everything you need to know about the program generating headlines across the globe.

## *Where is Artemis 1 going?*

The 42-day Artemis 1 mission will test the Orion spacecraft, a capsule that will circle the moon and one day take human crew members there. The uncrewed mission will launch from Cape Canaveral, Florida, no sooner than August 29 at 8:33 a.m. Eastern time, with September 2 and September 5 as backup options. Once in the atmosphere, Orion will begin in Earth's orbit, then fly into space fueled by the Interim Cryogenic Propulsion Stage (ICPS), a 45-foot long cylindrical structure with one engine. As Orion sails near the moon, a service module supplied by the European Space Agency will course-correct as required. The spacecraft will complete up to one and a half rotations in lunar orbit, where it will establish a record for the furthest any spacecraft that can carry a crew has flown. Then, it'll start its engines at exactly the perfect moment to be driven back toward Earth, with the assistance of the moon's gravity. On October 10, the Orion spacecraft will make a thundering return to our atmosphere it will be traveling at 6.8 miles per second, the fastest reentry of any capsule intended for people. The vehicle and its heat shield will have to sustain temperatures of 5,000 degrees Fahrenheit a vital component of this test trip because NASA can't intentionally produce these conditions on the ground, says Gizmodo's George Dvorsky. If it survives, Orion will splash down in the Pacific Ocean off the coast of San Diego, within view of a U.S. Navy ship that will rescue the spacecraft.

## *What's Remarkable About The Mission's Rocket, Dubbed The Space Launch System?*

The SLS is the most powerful rocket ever constructed, period. It rises to 32 floors tall and weighs over 6 million pounds. To construct it, NASA engaged various firms Northrop Grumman worked on the boosters, Aerojet Rocketdyne developed the engines, and Boeing manufactured the rocket's orange core stage. The project cost around $23.8 billion, a number that garnered considerable criticism for going over budget. When the SLS launches, it will be pushed by around 8.8 million pounds of force, a statistic that dwarfs the Saturn V rocket that launched the Apollo missions, which had 7.5 million pounds of thrust, Gizmodo writes. But when SpaceX's Starship, which is presently in construction, goes into flight, it will win the distinction of the most powerful rocket with its roaring 17 million pounds of thrust, planned to transport humans to deep-space destinations. Still, "SLS is the only rocket that can transport Orion, passengers and cargo straight to the Moon on a single mission," according to NASA.

## How Else Will This Flight Contribute To Science?

Though no person will ride aboard Artemis 1, three mannequins will go to outer space. Their mission: is to assess if the circumstances within the Orion spacecraft are safe for future astronaut passengers. At the head of the capsule will be Commander Moonikin Campos, a test dummy donning the Orion Crew Survival System spacesuit, according to Insider's Paola Rosa-Aquino. Sensors will assess the acceleration, vibration, and radiation that Moonkin is subjected to, providing NASA insight into how its human crew members could suffer. The other two mannequins, called Zohar and Helga, will assess how space radiation affects a woman's body. The dummies are built with slices of plastic that imitate soft tissue, bones, and lungs. Each will feature 5,600 sensors that will capture information on radiation's impact on the lungs, stomach, uterus, and bone marrow. Zohar will wear a safety vest, while Helga will not. As NASA prepares to send the first woman to the moon, this study is vital. "Women, in general, have a greater chance of acquiring cancer as they have more radiation-sensitive organs like breast tissue and ovaries," Ramona Gaza, scientific team lead at NASA's Johnson Space Center, said in a press conference. Artemis 1 will also carry 10 CubeSats or shoebox-sized satellites that commonly include items for study. The ICPS, after providing Orion its first push into space, will next disconnect from the spaceship and release these satellites at three distinct points between Earth and the moon. One of these CubeSats will utilize a solar sail to push it to a near-Earth asteroid, which it will image. Another includes yeast to assess how space radiation affects live cells. The other CubeSats will examine lunar ice with a spectrometer, picture the moon and the spaceship, test airbags in a lunar crash landing, and explore other scientific concerns.

## *Why Has This Expedition Been So Delayed?*

Ahead of the first flight test, the SLS rests on the launch pad at Cape Canaveral, Florida. NASA / Keegan Barber. Artemis 1 had previously been slated for a 2016 launch, according to the Orlando Sentinel's Richard Tribou. But a variety of variables complicated and delayed this aim, NASA administrator Bill Nelson said at a media conference last year. Manufacturing delays for both SLS and Orion, the Covid-19 outbreak, and trouble securing enough financing from Congress all rendered this date infeasible. This year, NASA struggled with the SLS rocket's wet dress rehearsals, or practice runs, leading up to Monday's launch. In April, the rocket failed three wet dress rehearsal efforts. Various difficulties, including a defective vent valve and a hydrogen leak, stopped NASA from finishing each test, as FLYING magazine's Jeremy Kariuki reported. The fourth effort in June ultimately worked: NASA filled the rocket's fuel tanks and proceeded through a countdown of 10 minutes before launch, until T-29 seconds. Despite another hydrogen leak that held up part of the rehearsal, NASA deemed the test a success.

## *What Are The Next Steps?*

Artemis 1 is to be followed by Artemis 2 and Artemis 3, missions that will culminate with astronauts once again walking on the moon. After this initial test flight, Artemis 2 will bring a human crew on a lunar flyby, entering the moon's orbit and returning in eight to ten days. Currently, the mission is slated for launch in 2024. If all goes according to plan, Artemis 3 will take place as soon as 2025. This mission will send a crew of astronauts to the moon's surface for the first time in more than 50 years. Last week, NASA identified 13 probable lunar landing places for the Artemis 3 crew to investigate, says Space.com's Meghan Bartels. All are near the lunar south pole, a region that scientists are targeting for investigation. In the arctic region's persistently dark and frigid climate, experts think frozen water may be discovered under the surface. As for which of these sites will be the destination, it will depend on the debut date. The Artemis program is only the beginning of NASA's "Moon to Mars" plan the agency hopes to make the moon a rest stop that will sustain people on longer space voyages. Artemis will set up the lunar Gateway, an outpost circling the moon that is to be constructed in space and facilitate future exploration. NASA also aims to put up a lunar base camp where astronauts may remain for long-term trips and test exploration technologies that might be utilized on Mars. By building on Artemis's successes, humans might be strolling on the Red Planet within 20 years. "Everything that we're doing on the lunar surface, we're doing to explore for science," Cathy Koerner, a deputy associate administrator at NASA, tells WIRED's, Ramin Skibba. "We're going not just for 'flags and footprints,' as some people refer to [Apollo], but also to test out all of the systems that we'll eventually need to bring down risks for a human mission to Mars."

# CHAPTER 4

## *IS ARTEMIS 1 MANNED?*

NASA is currently conducting many space missions as part of the Artemis program. The first of three Artemis missions, Artemis 1, will launch on August 29, 2022, and serve as an unmanned test flight around and beyond the Moon. Humanity will fly farther than ever in space on Artemis 2, a crewed mission beyond the Moon. And Artemis 3, a mission that will send the first female and the first person of color to the Moon to conduct research on the lunar surface for a week. Since Apollo 17 in 1972, Artemis 3 will be the US space agency's first crewed Moon landing mission. While NASA's long-term objectives are somewhat more ambitious, the Artemis space missions are primarily focused on lunar exploration. NASA plans to launch a crewed mission to Mars in the future using the technology and information acquired during the Artemis spaceflights. The goal of NASA's ambitious "Moon to Mars" plan is to eventually establish a habitable Moon colony as well as a new space station in lunar orbit.

## Why Is Artemis The Name Of The Program?

Apollo's twin sister and the fabled Greek moon goddess Artemis. It is thus obvious that there is a connection to the trip that was the first to send humans to the Moon fifty years ago. The crewed spacecraft that is currently under production is known as Orion. One of the most well-known constellations in the sky is Orion, and in classical mythology, Artemis's hunting companion.

# Information On The Artemis Mission

*2017 to the present*
*Vehicles for launch: Space Launch System (SLS); Private launch vehicles*
*Lunar Gateway, Orion, and the Human Landing System (HLS)*

## *Release Dates:*

- The launch of Artemis 1 is scheduled for 29 August 2022.
- Artemis 2: 2024 or later (TBC) (TBC)
- Aphrodite 3: 2025

## *Why Is Nasa Visiting The Moon Once More?*

NASA is not only trying to recreate the exploits of the Apollo missions with Artemis, but rather to get to the Moon 'and remain there. That entails researching the prospect of creating outposts both in lunar orbit and on the Moon's surface, while the major aim for now still involves sending people to the Moon by the middle of the decade.

## *Key Nasa Mission Goals Include:*

- Equality: a primary target for NASA is to place the first woman and first person of color on the lunar surface.
- Technology: from rockets to spacesuits, the technologies now being researched are aimed to pave the path for future deep-space expeditions.
- Partnerships: the Artemis project is one of NASA's first large-scale partnerships with private businesses, including SpaceX and Boeing.
- Long-term presence: while the Apollo 17 crew spent three days on the lunar surface, Artemis seeks to create a base

to prolong the excursions to weeks and potentially months.

- Knowledge: while more is known about the Moon compared with 50 years ago (and technology has substantially evolved), NASA says that this next series of missions will be able to gather samples more strategically than during the Apollo period.
- Resources: the finding of water on the Moon and probable reserves of rare minerals offer promise for both scientific and commercial research and utilization.

## Has The Artemis Launch Been Delayed?

NASA's first stated objective was to put people on the Moon by 2024. The agency said in November 2021 however that this date will be moved out to no sooner than 2025. Even this timetable is far from guaranteed, with NASA Inspector General

Paul Martin indicating that the crewed lunar landing would likely slide until 2026 at the earliest. The landing is the third of a multi-year mission commencing with Artemis I, an uncrewed voyage around and beyond the Moon. Each of these missions has ambitious deadlines of its own.

## *How Will Nasa Go Back To The Moon?*

There are four key components of the Artemis Moon missions. These include:

**Orion Spaceship**
Equipped with life support systems and shuttle interfaces, Orion is the command module required to convey the crew across space.

**Lunar Gateway**
The Lunar Gateway is a tiny space station circling the Moon, intended to provide a versatile platform for missions to the Moon

and beyond. The Orion module will connect with Gateway, and from here the crew will transfer into the lunar landing module. Unlike the International Space Station (ISS), the Lunar Gateway won't be permanently manned, but will serve as a platform where astronauts may dwell and do research for brief durations. It will also be able to continue scientific study even between human lunar expeditions. International partners like the European Space Agency are cooperating with NASA on the design of the Lunar Gateway.

## Moon Landing Module

The lunar landing vehicles will transfer freight and personnel from the Lunar Gateway to the Moon's surface. NASA is collaborating with private businesses to create both a human landing system (known as HLS) and several additional vehicles for robots and freight. Where Apollo's Lunar Module was planned to be utilized for one return flight to the Moon's surface, the landing systems for the Artemis missions are set to be used for several missions.

**Space Launch System (Sls)**

Tying together all these pieces is the rocket that will take them beyond Earth's atmosphere and into space. This super heavy-lift rocket is higher than the Statue of Liberty at 322ft and is predicted to cost $800 million per launch. When complete, the SLS will be the most powerful rocket in the world, more powerful than the original Saturn V launcher that first transported people to the Moon. The launcher has been under development at NASA for much of the past decade, facing several delays and growing costs. In March 2022, the rocket and spacecraft for Artemis I was transported to the launch pad at Kennedy Space Center.

# Will Traveling To The Moon Help Humanity Settle On Mars?

While the voyage to the Moon takes three days, reaching Mars is a significantly longer and more challenging endeavor. NASA envisions Artemis as establishing the basis for both international space agencies and private firms to construct a lunar population and economy, and from there someday take people to Mars.

## NASA Artemis Mission Timetable

### Artemis 1

Formerly designated Exploration Mission-1, this uncrewed mission is a comprehensive test of the Space Launch System (SLS) and the Orion module. The SLS will blast off from the Kennedy Orbit Centre in Florida, and once in orbit, the Orion module will detach and continue to the Moon. Its orbit will take it 100 kilometers above the lunar surface before it continues around 64,373 miles beyond the Moon. The module is anticipated to last 42 days and will splash down in the Pacific Ocean not far from California. Numerous technical and scientific investigations that take place in deep space will be conducted on board. These are intended to boost technological development, increase our knowledge of the moon, and provide information about radiation in deep space. Campos, a mannequin wearing the First-Generation Orion Crew Survival System spacesuit that the real astronauts will wear on Artemis 2 and 3, will also be on board. Sensors will be installed all around Campos to provide information about potential flying experiences for human crew members.

## 2 Artemis

The Artemis project will be a ground-breaking crewed spaceflight that will take humans further than they've ever been in space. The four-person crew will fly the Orion module 7402 kilometers beyond the far side of the Moon after being propelled into orbit by the SLS rocket, perform a lunar flyby, and then come back to Earth. The expedition will gather crucial flight test data over eight to ten days.

## 3 Artemis

After Apollo 17 in 1972, the third mission to the Moon is scheduled to provide the first Moon landing ever. Four astronauts on board the Orion module will continue the legacy of the Artemis 2 mission by docking with the Lunar Gateway and spending 30 days in orbit. The two astronauts will then be transported by the human landing system to the South Pole of the Moon, a location that has never been explored by humans. The astronauts will spend a week on the surface exploring and conducting a range of scientific investigations, including collecting water ice, which was discovered for the first time on the Moon in 1971.

## 4, 5, 6, And More Artemis?

In addition to looking forward to future projects, NASA is largely concentrating on Artemis missions 1 through 3, but they have also already granted contracts for boosters on rockets up to Artemis 13! NASA hopes to launch further crewed flights each year if missions 1-3 are successful. Future astronauts are anticipated to start a station on the Moon's surface to utilize the satellite as a staging area for missions to Mars. As of January 2022, NASA will publicly announce the casting and assembly of Artemis 4's solid rocket booster. In the meanwhile, Airbus announced in June 2022 that they had received the module's structural components in time for the fourth Artemis mission, which would allow humans to reside in lunar orbit.

## *Who Will Be On Board The Artemis?*

NASA began its search for new astronauts in 2020 intending to send people to the Moon, Mars, and beyond. Each applicant has to be a citizen of the United States and have a master's degree in a STEM discipline (science, technology, engineering, or maths). The application period for the next intake ended on March 31, 2020, and the new "Artemis Team" was unveiled in December of that year. However, NASA said in August 2022 that all 42 of its astronauts—not only the 18 members of the "Artemis Team"—would be qualified to fly. The crew for Artemis 2 is anticipated to be chosen shortly. Shaun the Sheep, who will be traveling further than any human or sheep before, will be aboard Artemis 1 even though it won't be crewed.

# CHAPTER 5

## *WHERE AND WHEN TO WATCH NASA LAUNCH*

The Artemis-1 mission will launch on Monday, August 29 within a two-hour window that begins at 8:33 a.m. EDT, according to a flight readiness assessment this week. If it doesn't launch on time, the next opportunity will be at noon on Friday, September 2, 2022. The Artemis-1 mission is unique in every way. a protracted journey there, far beyond it, and back. Before crewed trips to the Moon, Artemis I, the first in a complicated series of missions, will test NASA's new heavy-lift rocket, the Space Launch System (SLS), and the Orion spacecraft as an integrated system. The SLS rocket, which stands 322 feet tall, will be the most powerful launch vehicle ever since NASA's last Saturn V "Moon rocket" placed the Skylab space station in Earth orbit in

1973. The SLS rocket has a thrust of 8.8 million pounds (3.9 million kg), making it the most powerful rocket ever built. This is something you must not miss!

## *Where Will Artemis-1 Take Off?*

Launch Pad 39B at the NASA Kennedy Space Center in Cape Canaveral, Florida, will serve as the launch site for Artemis-1. The launch of Artemis-1 is planned for Monday, August 29, 2022. The launch window is available from 8:33 a.m. until 10:33 a.m. EST. There are two other launch dates if the planned launch is postponed: Friday, September 2 (12:48–14:48 EDT), and Monday, September 5. (17:12 - 18:42 EDT).

## The Reason Why Artemis-1 Can't Launch A Day Later

The Moon's location is in the sky. The solar-powered Orion spacecraft's course must avoid the Moon's shadow for more than 90 minutes if it does so. If it does, it will lose all power. The next launch window is September 19, 2022, if it doesn't take off by September 6.

## When And Where To See The Launch Of Artemis-1

NASA On Monday, August 29, 2022, YouTube will have full coverage of the launch. Although programming starts at midnight, launch coverage begins at 6:30 a.m. Additionally, NASA TV is accessible through Facebook, Twitch, the NASA website, and in 4K on the NASA UHD channel. The full timetable and certain activities you shouldn't miss in the hours after launch are listed below:

English-language live launch coverage starts at 6:30 a.m. (Spanish is 7:30 a.m. EST on separate feeds on Twitter, Facebook, and YouTube).

## *Launch Window: 8:33–10:33 Est.*

A post-launch press conference will take place an hour or so after liftoff.
Coverage of Orion's first outbound trajectory burns on its route to the Moon at 4:00 p.m. EST. The precise liftoff time determines the exact time.
First Earth sights from Orion on the outgoing coast to the Moon are covered at 5:30 EST. The precise liftoff time determines the exact time.

## *Where To Obtain Updates On The Launch Of Artemis-1*

Be ready for delays, scrubs, etc. since rocket launches seldom occur on schedule. Plan, but be aware that you'll probably need to change your plans. For updates on launch schedules and coverage modifications, keep a watch on the @NASAArtemis and @NASA Twitter accounts. After launching and orbiting the Earth, the spacecraft and rocket will take Orion and the ESM into an elliptical orbit around the Moon, where they will pass within 69 miles (111 kilometers) of the surface around September 7, 2022, and then roughly 40,000 miles beyond it in the days that follow. That is farther than any human-built spacecraft has ever traveled. On its route back to Earth, it will then loop around and fly even closer to the Moon. The Artemis-1 mission, assuming a successful launch, will span 42 days and splash down in the Pacific Ocean off the coast of San Diego, California, on October 10, 2022.

## *The Final Objective Of The Artemis Program*

The launch of Artemis-1 will signal the start of a human return to the Moon. The Apollo 17 astronauts Jack Schmitt and Gene Cernan were the last people to set foot on the moon in 1972. The Artemis-1 mission is expected to bring the first woman and the first person of color to the Moon's south pole in 2025. The next mission, Artemis-2, will launch in 2024 and will essentially be a repetition of Artemis-1 with four people on board to test Orion's life support systems. The 10-day Artemis-2 mission will begin with two Earth orbits and end 4,600 kilometers beyond the far side of the Moon.

# CHAPTER 6

## WHY DIDN'T WE VISIT THE MOON AGAIN?

In the course of the Apollo project, the U.S. National Aeronautics and Space Administration (NASA) sent 12 astronauts to the Moon between 1969 and 1972. However, despite multiple successive presidential policy efforts, nobody has set foot on the Moon in the years that have followed. For the United States, the Apollo program was an expensive undertaking. While historical sources differ on the price of the program, most believe that it cost at least $20 billion in 1973 dollars (about $116 billion in 2019). Compared to previous years, NASA has used around 0.5 percent of yearly government expenditure, down from nearly 4 percent at its height in the middle of the 1960s.

# Cost Of The Apollo Program In 1973 Compared To Its Equivalent Cost In 2019

Apollo 20 was to be used by NASA to launch human trips to the Moon, while the Apollo Applications Program was to convert Moon mission technologies for use on future exploration missions (AAP). However, NASA funding reductions by Congress hastened the termination of the Moon mission until Apollo 17, in 1972. Except for the space station Skylab, the majority of AAP projects were abandoned. Congress cut NASA's budget for several reasons. The space race, a battle between the Soviet Union and the United States to demonstrate their military and technical supremacy to other countries, served as the primary push for the Moon mission. The competitive spirit cooled to détente later in the 1960s, however, and the strategic imperative of funding NASA diminished. Other public objectives were also beginning to emerge, the most important of which was the costly Vietnam War, which demanded a sizable portion of government funding. After Apollo 11, on July 20, 1969, the first human Moon landing, public interest in space likewise declined. In their 1997 book Spaceflight and the Myth of Presidential Leadership, space historians Roger D. Launius and Howard E. McCurdy make a further case that Apollo was born out of a special situation. Because he was worried about Soviet military might, U.S. President John F. Kennedy specifically pushed the space program and the Moon landings as one of the country's main priorities. NASA and its projects transitioned to an auxiliary policy with détente and have stayed there ever since. NASA's goals shifted throughout the next decades following legislative demands, and its increasingly constrained funds for human spaceflight were allocated to initiatives other than moon exploration. The largely reusable Space Shuttle, whose five spacecraft completed 135 trips between 1981 and 2011, was the next significant project following Apollo. Additionally, NASA worked on several space station designs that ultimately led to its participation in the International Space Station (ISS), whose first components were launched in 1998. The International Space Station (ISS) was promoted as a platform for global policymaking and as a research laboratory, particularly about Russia, which at the time was a young country emerging from the demise of the Soviet Union. Over the years, additional Moon endeavors have been presented by three presidents, but the majority of these plans were shelved owing to financing issues and decreasing legislative support. These were George W. Bush's Vision for Space Exploration, which called for Moon missions by 2020, and George H.W. Bush's Space Exploration Initiative, which aimed to place people on the moon by the turn of the century. Both programs were abandoned soon after each president's terms in office. The Gateway lunar space station and Project Artemis, which aims for manned landings by the year 2024, are two significant Moon endeavors currently being planned by the Donald Trump administration. According to Jim Bridenstine, the administrator of NASA, the fresh Moon landings planned for Project Artemis may cost the agency between $20 billion and $30 billion in 2019 dollars. Compared to the estimated $115 billion price tag for Apollo, this would be far less expensive.

## THE COST OF PROJECT ARTEMIS COULD RUN FROM $20 BILLION TO $30 BILLION.

No country in the 1960s, outside the United States and the Soviet Union, had space programs developed enough to contemplate sending people to the Moon. However, in recent years, nations such as China, India, Japan, Russia, and those that are a part of the European Space Agency have all made public assumptions about potential Moon landings. For the Artemis and Gateway partnerships, NASA is looking for ISS partners. Canada has agreed to contribute robots to the Gateway, making it the sole partner that has committed as of this writing. Any nation or organization that decides to send astronauts to the moon will have to assume some risk and make financial commitments. Since people need water, air, food, and other comforts to stay alive, human moon missions take more resources than robotic moon landings. However, several countries, including private businesses from those countries, are developing robotic Moon ventures that might help future manned expeditions.

9 798848 593310